NATIONAL
GEOGRAPHIC
KiDS

美 国 国 家 地 理
双 语 阅 读

Abraham
Lincoln

亚伯拉罕·林肯

懿海文化 编著

马鸣 译

第三级

外语教学与研究出版社
FOREIGN LANGUAGE TEACHING AND RESEARCH PRESS
北京 BEIJING

京权图字：01-2021-5130

图书在版编目（CIP）数据

亚伯拉罕·林肯：英文、汉文／懿海文化编著；马鸣译. —— 北京：外语教学与研究出版社，2021.11（2023.8 重印）
（美国国家地理双语阅读. 第三级）
书名原文：Abraham Lincoln
ISBN 978-7-5213-3147-9

Ⅰ. ①亚… Ⅱ. ①懿… ②马… Ⅲ. ①林肯(Lincoln, Abraham 1809−1865)－传记－少儿读物－英、汉 Ⅳ. ①K837.127=41

中国版本图书馆 CIP 数据核字 (2021) 第 228169 号

出 版 人　王　芳
策划编辑　许海峰　刘秀玲　姚　璐
责任编辑　姚　璐
责任校对　华　蕾
装帧设计　许　岚
出版发行　外语教学与研究出版社
社　　址　北京市西三环北路 19 号（100089）
网　　址　https://www.fltrp.com
印　　刷　天津海顺印业包装有限公司
开　　本　650×980　1/16
印　　张　37.5
版　　次　2022 年 3 月第 1 版 2023 年 8 月第 4 次印刷
书　　号　ISBN 978-7-5213-3147-9
定　　价　188.00 元（全 15 册）

如有图书采购需求，图书内容或印刷装订等问题，侵权、盗版书籍等线索，请拨打以下电话或关注官方服务号：
客服电话: 400 898 7008
官方服务号: 微信搜索并关注公众号"外研社官方服务号"
外研社购书网址: https://fltrp.tmall.com

物料号: 331470001

Table of Contents

A Much-Loved President 4

A Country Boy 6

Growing Up 8

In His Time 10

Becoming a Leader 12

9 Awesome Facts 14

Slavery 16

Civil War 20

The Gettysburg Address 24

Lincoln's Last Days 26

Lincoln's Life 28

Quiz Whiz 30

Glossary 32

参考译文 33

A Much-Loved President

Who's on the penny? And the five-dollar bill? It's Abraham Lincoln, the 16th President of the United States!

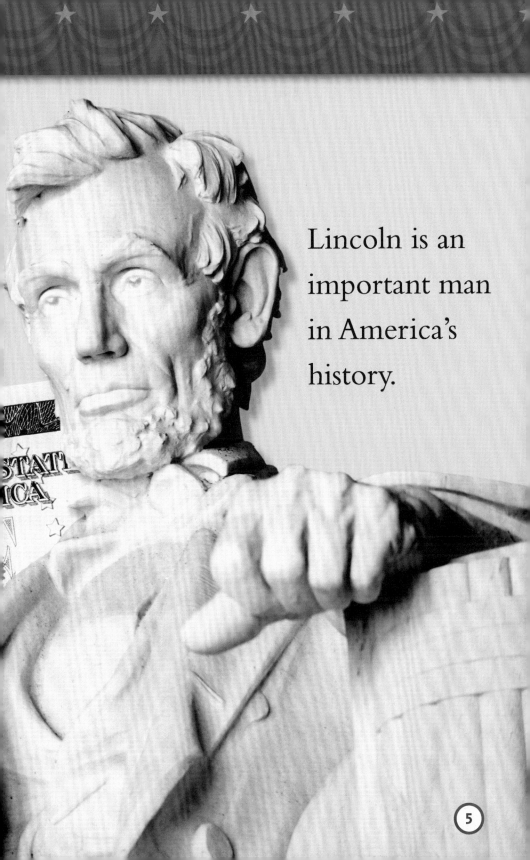

Lincoln is an important man in America's history.

A Country Boy

Lincoln was born in a one-room log cabin in Kentucky on February 12, 1809. He grew up in Indiana.

Lincoln read a lot of books.

Lincoln's family home in Kentucky

He wanted to go to school, but there was too much work to do on his family's farm. He chopped down trees, built fences, and plowed the land. He grew strong and very tall. Lincoln was kind and a good storyteller.

In His Own Words

"Leave nothing for tomorrow which can be done today."

Growing Up

All his life, Lincoln taught himself how to do things. He learned how to read and write, tell stories, and give speeches.

That's a Fact! Lincoln was an inventor. He had an idea for a machine to help ships float over sandbars.

LINCOLN - HERNDON LAW OFFICES

Word to Know

LAWYER: A person who provides advice about the law

Lincoln learned how to fix machines. He taught himself how to pilot a riverboat and how to be a soldier. He even studied law and became a lawyer—all on his own.

In His Time

In the 1800s, many things were different from how they are today.

Money

A pair of shoes cost one dollar. A quart of milk cost ten cents. That doesn't sound like much, but dollars and dimes were worth a lot more back then.

Toys

In their free time, children played hopscotch and leapfrog. They also played with marbles, dolls, and toy trains.

School

Back then, not all children went to school. Those who did learned together in a one-room schoolhouse.

Transportation

People walked and traveled by horse-drawn carriages. Trains were used for long trips. There were no cars or airplanes.

U. S. Events

California became the 31st state in 1850. *The New York Times* newspaper started in 1851.

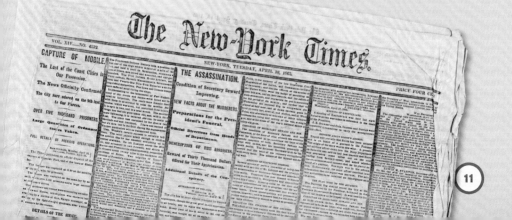

Becoming a Leader

In 1842, Lincoln married Mary Todd. They had four sons, named Robert, Eddie, Willie, and Tad.

A painting of the Lincolns with three of their sons, Willie, Robert, and Tad

That's a Fact!

Mary Todd was one of fifteen children in her family.

Lincoln was a good lawyer. People trusted him to make important decisions. So Lincoln became a politician (pall-uh-TISH-un).

Lincoln held two jobs as a politician for the state of Illinois. Then, in 1860, he ran for President. Lincoln won! He became the 16th President of the United States.

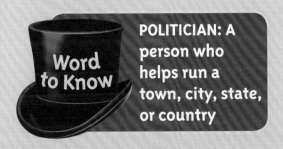

Word to Know

POLITICIAN: A person who helps run a town, city, state, or country

9 Awesome Facts

2 The Lincoln penny has looked the same since 1909, with Lincoln on the "heads" side.

1 Eleven-year-old Grace Bedell wrote Lincoln a letter suggesting he grow a beard—and he did!

3 Lincoln made Thanksgiving a national holiday in 1863. Magazine editor Sarah Josepha Hale suggested the idea to Lincoln.

4 Lincoln read the Bible often but did not belong to any church.

5

Lincoln owned horses, cats, dogs, and a turkey. Once, he saved his dog from drowning in a frozen river.

6

Some Presidents use speechwriters, but Lincoln wrote all his own speeches.

7

Lincoln's stepmother was an important person in his life. She encouraged him to read and learn.

8

Mary Todd Lincoln was more than a foot shorter than her husband.

9

Lincoln kept important papers in his stovepipe hat. This kept his head warm and his papers dry on rainy days.

Slavery

During Lincoln's time, some white people owned black slaves. The first slaves were brought to America from Africa. They were stolen from their families, and sold to people who used them to do work.

In His Own Words

"I want every man to have the chance—and I believe a black man is entitled to it—in which he can better his condition."

Slaves did not get paid and had to obey their owners. Slaves did not have any rights. They were often treated very badly. Many people were slaves because their parents were slaves. Those who tried to escape were often caught. Most slaves would never be free.

100 DOLLS. REWARD.
RAN AWAY
From me, on Saturday, the 19th inst., Negro Boy Robert Porter, aged 19; heavy, stoutly made dark chesnut complexion sullen countenance

An illustration of slaves working on a southern farm shows the hard work needed to grow and harvest crops.

People all over the country disagreed about slavery. Many people in the South wanted slaves to work on their large farms. Most people in the North worked in cities and thought slavery was wrong.

That's a Fact!

Slaves used music for comfort and support. They sang about freedom, hard work, and their beliefs.

Lincoln was against slavery, and told people it needed to end.

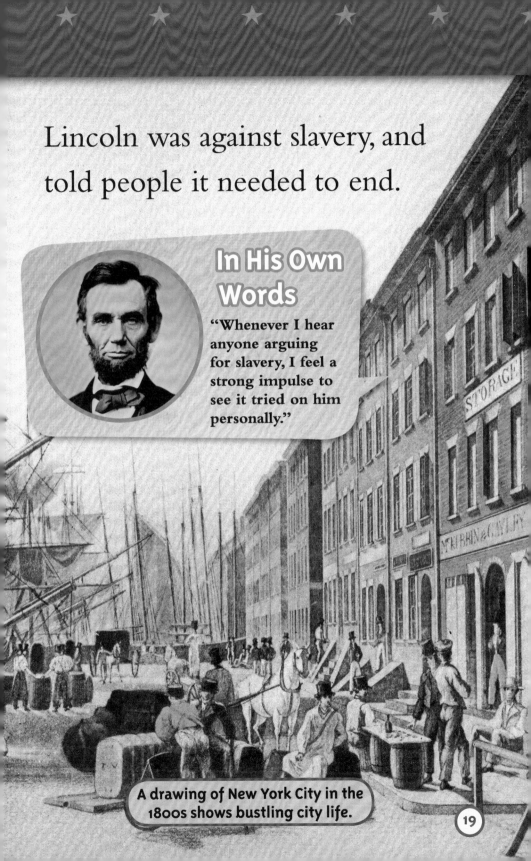

In His Own Words

"Whenever I hear anyone arguing for slavery, I feel a strong impulse to see it tried on him personally."

STORAGE

A drawing of New York City in the 1800s shows bustling city life.

Civil War

In 1861, 11 southern states broke away from the United States, which was also called the Union. These states did not want to be part of the Union because most of the Union wanted slavery to end.

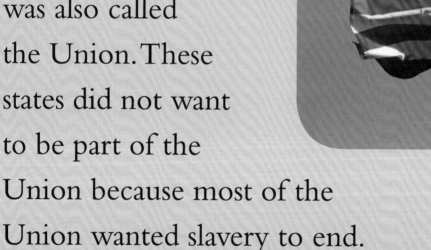

The country began a long and bloody Civil War.

Words to Know

CIVIL WAR: A war between people of the same country

THE UNION: The United States, but only northern states during the Civil War

Families disagreed and broke apart. Sometimes brothers fought on opposite sides. Some slaves escaped to the North to fight for freedom. Many people on both sides died.

In 1863, Lincoln freed slaves in ten states, but the war continued. Months later, he gave his most famous speech: the Gettysburg (GET-tees-burg) Address.

In His Own Words

"I say 'try'; if we never try, we shall never succeed..."

The Gettysburg Address

Abraham Lincoln stood on the battlefield where thousands had died to end slavery. He said that the country began with the idea that all people should be free. People listened to Lincoln's powerful words.

Abraham Lincoln at Gettysburg

In His Own Words

"Four score and seven years ago our fathers brought forth on this continent, a new nation, conceived in Liberty, and dedicated to the proposition that all men are created equal. . ."

—*The Gettysburg Address*

In 1865, after many more battles, the South surrendered. The North had won. The long, terrible war was over.

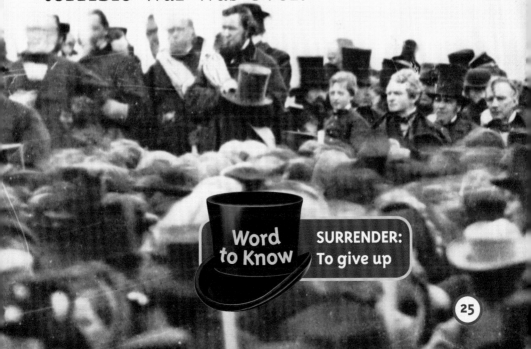

Word to Know

SURRENDER: To give up

Lincoln's Last Days

After the war, some Southerners were still angry. They didn't want Lincoln as President. A few were so angry they wanted to assassinate Lincoln.

On April 14, 1865, Lincoln went to the theater. During the play, he was shot by a man named John Wilkes Booth. Lincoln died the next day. Booth got away but was found and killed for his crime. Americans were sad to have lost Lincoln, their leader.

Word to Know

ASSASSINATE: To murder an important person

Lincoln's Life

Lincoln left behind a free country. The Union had been saved. Over time, Lincoln became one of the most loved of all American Presidents.

Each year, many people visit the Lincoln Memorial in Washington, D. C., and the Lincoln Presidential Library in Springfield, Illinois.

1809	1816	1830	1834	1842
Born in Kentucky to Thomas and Nancy Lincoln	Moved to Indiana	Moved to Illinois	Elected to Illinois State Legislature	Married Mary Todd

Lincoln Memorial

When Barack Obama was sworn in as America's first black President, he used the same Bible that Lincoln used.

★ ★

1846
Elected to U.S. House of Representatives

1860
Elected President of the United States

1863
Freed the slaves in ten states; gave the Gettysburg Address

1864
Elected President for a second time

1865
Died in Washington, D. C., on April 15

Quiz Whiz

See how many Lincoln questions you can get right! Answers are at the bottom of page 31.

1

Lincoln was the ____ President of the United States.
A. 4th
B. 16th
C. 10th
D. 1st

2

In Lincoln's time, _____.
A. Not all children went to school
B. People traveled by plane
C. People watched TV
D. Shoes cost ten dollars

3

What was Lincoln's wife's name?
A. Mary McDonald Lincoln
B. Marion Lincoln
C. Mary Todd Lincoln
D. Mary Adams

4

The night Lincoln was shot, he was at _____.
A. The White House
B. The theater
C. His family's farm
D. His law office

A slave _____.
A. Had to obey his or her owners
B. Was not paid for his or her work
C. Usually worked on large farms
D. All of the above

5

6

What is a war between people of the same country called?
A. A treaty
B. A civil war
C. A disagreement
D. A skirmish

How many sons did Lincoln have?
A. Six
B. Four
C. None
D. One

7

Glossary

ASSASSINATE: To murder an important person

CIVIL WAR: A war between people of the same country

LAWYER: A person who provides advice about the law

POLITICIAN: A person who helps run a town, city, state, or country

SURRENDER: To give up

THE UNION: The United States, but only northern states during the Civil War

▶ 第4—5页

备受爱戴的总统

一美分硬币上的人是谁？五美元纸币上的人又是谁呢？他是亚伯拉罕·林肯，美国的第16任总统。

在美国的历史上，林肯是个重要人物。

▶ 第6—7页

一个乡下男孩

1809年2月12日，林肯出生在肯塔基州一座只有一个房间的小木屋里。他在印第安纳州长大。

林肯读了很多书。

他想去上学，但他家的农场上有很多活儿要做。他砍树，搭栅栏，犁地。他长得又壮又高。林肯很善良，还很会讲故事。

林肯语录

"今日事，今日毕。"

林肯在肯塔基州的老家

▶ 第8—9页

长大成人

终其一生，林肯都自学怎么做每一件事。他学会了读写、讲故事和演讲。

林肯学会了如何修理机器。他自学如何驾驶汽船以及如何成为一名士兵。他甚至学习法律，成为一名律师——全靠自学。

不可思议的事实

林肯是个发明家。他曾想改良一台机器，用来帮助船穿过沙洲。

小词典

律师：提供法律建议的人

33

他所在的时代

在 19 世纪，很多东西都和它们现在的样子不一样。

货币

一双鞋卖1美元，1夸脱牛奶（约0.95升）卖10美分。这听起来并不贵，但在那个时候，美元和1角镍币能买到更多东西。

玩具

孩子们在闲暇时玩跳房子和跳背游戏。他们还玩弹珠、娃娃和玩具火车。

学校

那个时候，不是所有的孩子都去上学。去上学的孩子们也只能挤在只有一间教室的学校里学习。

交通

人们走路及坐马车出行，长途旅行时会坐火车。那个时候没有汽车，也没有飞机。

美国大事件

1850年，加利福尼亚州成为美国的第31个州。1851年，《纽约时报》正式发行。

成为领袖

1842 年，林肯和玛丽·托德结婚了。他们生了四个儿子，名叫罗伯特、爱德华、威廉和塔德。

林肯是一名优秀的律师。人们信任他，请他做重要决定。因此，林肯成了一名政治家。

作为政治家，林肯在伊利诺伊州有两份工作。接着，1860 年，他竞选美国总统。林肯成功了！他成为美国第 16 任总统。

林肯夫妇和三个儿子威廉、罗伯特和塔德的画像

不可思议的事实

玛丽·托德是她家15个孩子中的一个！

小词典

政治家：帮助某个小镇、城市、州或者国家运转的人

9 个意想不到的事实

1 11岁的格雷斯·贝德尔给林肯写信，建议他留胡子——他真的照做了！

2 1909年开始发行的一美分硬币从未变过，正面始终是林肯的头像。

3 1863年，林肯将感恩节定为全国性节日。这个主意是一个叫萨拉·约瑟法·赫尔的杂志社编辑向林肯提出的。

4 林肯经常读《圣经》，但他没有加入任何教派。

5 林肯养马、猫、狗，还养过一只火鸡。他曾经从冰冷的河里救了他养的一条狗，免得它被淹死。

6 有的总统会找人代写演讲稿，但林肯的演讲稿都是他自己写的。

7 林肯的继母在他的生命中非常重要。她鼓励他读书、学习。

8 玛丽·托德·林肯比她的丈夫矮1英尺（约30.48厘米）。

9 林肯把重要的文件放在他的大礼帽里。在下雨天，这能为他的头保暖，也能保证文件是干的。

▶ 第 16—17 页

奴隶制

在林肯的那个时代，一些白人拥有黑奴。第一批奴隶是从非洲被带到美国的。他们被人从家里偷出来，卖给那些用他们做工的人。

奴隶既得不到报酬，又必须服从他们的主人。奴隶没有任何权利。他们经常受到虐待。很多人之所以是奴隶，是因为他们的父母是奴隶。那些想要逃跑的奴隶往往都被抓回去。大多数奴隶一辈子都没有自由。

林肯语录

"我想让每个人都有机会改善自己的生活状态——我认为黑人也有这样的机会。"

▶ 第 18—19 页

关于奴隶制，全国人民看法不一。南方的很多人想让奴隶继续在他们的大农场上工作。北方的大多数人在城市里工作，他们认为奴隶制是不对的。

林肯反对奴隶制，告诉人们它该终结了。

奴隶在南方的农场上工作的图画展现了工作的辛苦，他们要种庄稼、收割庄稼。

林肯语录

"每当我听到有人为奴隶制辩解，我就有种冲动，想看看他当奴隶的样子。"

19世纪的纽约城的绘画展现了忙碌的城市生活。

不可思议的事实

奴隶们以音乐为慰藉、支撑。他们歌唱自由和辛苦的工作，以及他们的信仰。

▶ 第 20—21 页

内战

　　1861 年，11 个南方州脱离美国，即脱离联邦。这些州不愿意继续作为联邦的一部分，因为联邦的大多数都想要废除奴隶制。

　　这个国家开始了一场漫长又血腥的内战，即南北战争。

小词典

内战：同一个国家的人民之间的战争

联邦：美国，但在南北战争期间只包括北方各州

▶ 第 22—23 页

　　很多家庭因政见不同而分崩离析。有时候兄弟为不同的阵营而战。一些奴隶逃到北方，为自由而战。双方都有很多人丧生。

　　1863 年，林肯解放了 10 个州的奴隶，但是战争仍在继续。几个月后，他发表了他最著名的演讲——葛底斯堡演说。

林肯语录

　　"我说'试一试'。如果我们不去尝试，我们永远不会成功……"

▶ 第 24—25 页

葛底斯堡演说

　　亚伯拉罕·林肯站在战场上，成千上万的人在那里为终结奴隶制而失去了生命。他说，国家建立的初衷是人人皆自由。人们聆听着林肯铿锵有力的言辞。

　　1865 年，在经历了多场战役之后，南方投降了。北方获得了胜利。这场漫长又残酷的战争结束了。

林肯语录

　　"87 年以前，我们的先辈在这块大陆上建立了新的国家，它在争取自由中诞生，奉行人人生来平等这一信念……"
——《葛底斯堡演说》

亚伯拉罕·林肯在葛底斯堡

小词典　投降：认输

▶ 第 26—27 页

林肯的最后时光

战争结束之后，一些南方人仍然很愤怒。他们不想让林肯做总统。有几个人特别愤怒，以至于想暗杀林肯。

1865 年 4 月 14 日，林肯去剧院。在观看演出时，一个叫约翰·维尔克斯·布思的男人朝林肯开枪。第二天，林肯就去世了。布思逃走了，后来他被抓获并处以死刑。美国人民为失去他们的领袖林肯而异常悲痛。

小词典　暗杀：谋杀重要的人物

▶ 第 28—29 页

林肯的一生

林肯留下了一个自由的国家。联邦得以挽救。渐渐地，林肯成为最受爱戴的美国总统之一。

每年都有许多人参观位于华盛顿的林肯纪念堂和位于伊利诺伊州斯普林菲尔德的林肯总统图书馆。

美国第一位黑人总统巴拉克·奥巴马宣誓就职时，使用了林肯曾经用过的那本《圣经》。

不可思议的事实　林肯纪念堂的 36 根柱子代表林肯去世时联邦的 36 个州。

林肯纪念堂

1809 年	1816 年	1830 年	1834 年	1842 年
出生于肯塔基州的托马斯·林肯和南茜·林肯家中	搬到印第安纳州居住	搬到伊利诺伊州居住	当选为伊利诺伊州议员	与玛丽·托德结婚

1846 年	1860 年	1863 年	1864 年	1865 年
当选为美国国会众议院议员	当选为美国总统	解放 10 个州的奴隶；发表葛底斯堡演说	再次当选为美国总统	4 月 15 日于华盛顿去世

答题小能手

看看你能答对几个有关林肯的问题！答案在第 31 页下方。

1 林肯是美国第 _____ 任总统。
A.4　　B.16　　C.10　　D.1

2 在林肯那个时代，_____。
A. 不是所有的孩子都去上学　　B. 人们乘坐飞机旅行
C. 人们看电视　　　　　　　　D. 一双鞋卖十美元

3 林肯妻子的名字是什么？
A. 玛丽·麦克唐纳德·林肯　　B. 玛瑞恩·林肯
C. 玛丽·托德·林肯　　　　　D. 玛丽·亚当斯

4 被暗杀的那一晚，林肯在 _____。
A. 白宫　　　　　B. 剧院
C. 他家的农场　　D. 他的律师事务所

5 奴隶 _____。
A. 必须服从主人　　B. 工作时得不到报酬
C. 一般在大农场上工作　　D. 以上都是

6 同一个国家的人民之间的战争叫什么？
A. 条约　　　　B. 内战
C. 异议　　　　D. 冲突

7 林肯有几个儿子？
A.6　　B.4　　C.0　　D.1

词汇表

暗杀：谋杀重要的人物

内战：同一个国家的人民之间的战争

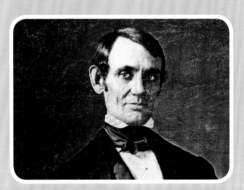

律师：提供法律建议的人

政治家：帮助某个小镇、城市、州或者国家运转的人

投降：认输

联邦：美国，但在南北战争期间只包括北方各州